HOW I PASSED THE PN & RN NCLEX-EXAMS

Faith & Hard Work

Petra Guillaume

Table of Contents

Poem: Flowers

Colors from heaven

Gifts for the soul

Jesus knew I would love flowers one of my favorite things to give

and to receive

Plants are up on my favorite list too especially when I am on a

budget

Flowers warm the heart and sing to the soul

What a sweet lullaby

Chapter 1

When I started in nursing I was at the age of 18. During my time at the long term facility where I worked as a unit clerk during the week day and as a CNA on weekends. Being a CNA during the earlier part of my career was hard work because of the direct care I provided for the residents. Some residents were total care meaning they could not participate in their care because they were unable to or choose not to. Most of the time the residents who were total care couldn't participate in their care because of their medical condition.

My goal was to become a teacher at first and I had just completed my associate degree in early childhood education when the nurses encouraged me to become a Registered Nurse at the long term facility where I worked. My caring personality and a great attitude allowed me to work with the residents with a great deal of patience. My major changed to nursing and I began the journey to becoming a registered nurse.

The day was going well and I had just finished my shift as a nurse on the psychiatric ward. The patients were as expected. Some were giving us a hard time others were too busy trying to cope with their situation. One young man came in and was fighting. He had to be medicated and retrained. Later in the day he showed us a behavior I will never forget. He looked as if someone or something went inside of him and wanted to get out but couldn't. He looked scary especially his throat and face area.

Besides being tired and exhausted on most days I was burned out and was preparing to exit psychiatry mentally at this point. There were other parts of nursing I could return to for example, I love to teach, so I could return as an educator in a facility or as a clinical instructor. With over fifteen years experienced as a nurse in various areas this decision was not difficult.

Chapter 2

My career definitely had a ladder type of order to it and I recommend such a move. Some days were challenging and other days were both challenging and rewarding. Thank God I stayed on course and returned each time to complete a degree higher up to accomplish my goals. The support system was in place and still is my family.

The memorable part of my journey was when I had to get up at 4 am every morning and travel to the Bronx where I met with my best friend who drove us to the upstate campus on Fridays and we will leave the campus on Sundays to return home. One day we had a terrible accident I truly thought we were going to die because the car spun out of control and ended up on a hill on the highway.

The state trooper thought maybe we were drinking. We were dedicated students in pursuit of our dreams who did not smoke or drink. Although he was doing his job we needed to be comforted at that particular time. We looked at each other and we kept silent. We knew we came close to losing our lives.

We had to complete the RN program and return home safely to our family and that was the mission and we completed the program with flying colors, pass the state board exam on the first attempt. The time had gone by rather quickly.

My best friend and I did everything we supposed to when came to the programs we attended. We had great memories along the way. The Licensed Practical Nursing program was successful as well. We were able to complete the LPN first then we went to complete the RN both with high grades and pass our state board exams on the first attempt, we were happy. Our families attended both graduations for the LPN and the RN programs.

Chapter 3

We made friends along the way and we were doing well. Another memorable event that took place during my RN AAS degree was in clinical for the first day I was given the largest patient on the unit. Politely I left the room after entering the room ready to work and zipped across the hallway to my best friend who couldn't believe her eyes. He was morbidly obese and I thought I was being punished. The nurses on the unit couldn't believe an instructor could do such a thing.

In the end it taught me to be mindful of the students and their abilities in the clinical area when I became a clinical instructor. Today I looked back and see how God worked out bad for good. Today I am more patient with students because of that particular experience.

Passing the NCLEX (National Council Licensure Examination) exams for the PN and the RN was not an easy task. I took some time off and went to the library every day and practiced questions from four to five NCLEX books. My entire focus was on passing my exams therefore I did not allow any distractions. The LPN give me a strong foundation for the RN material. The best decision I made when comes to passing my state exams was taking the needed time off. The class material I knew however, I had to practice how to take the exam. The type of questions I would come across and how to answer those particular questions. The class material transformed into a question format is essentially what the state exams were about and your job is to recognize the answers right away in that particular format. My entire focus was on the questions in the books and on the CDs. In the library I answered the questions in its entirety from page to page checking my answers once I was finish with a test. Pay particular attention to the rationale behind the answer and pay particular attention to the ones you are getting wrong for example, if you are getting pharmacology type of questions wrong then this is an opportunity to go back into your class notes and the book you are practicing from to brush up on pharmacology. Do not panic, stay calm. You got this. You prepared for this for the past two to four years. At home I practiced the questions on the CDs that came with the book. This is a great assimilation because the entire LPN and the RN state board exams are given in this format on a computer at a testing site. The night of the exam I did not study or practice any questions. The morning of the exam I had a light breakfast and went to the testing site to take my exam. I was placed in front of a computer and given some brief instructions and then I took my exam with students in the room. We each had our own computer with some distance between us and proctors in the room. Everyone had a different set of questions in a sequence. The computer generated questions based on how well you are doing. At the end of the exam I felt I did Ok. I received my results in the mail and I passed on the first attempt on both exams. I was ecstatic; my hard work had paid off. This is a simple way to prepare for the exam and it works. Good luck with the exams and remember to stay calm and be confident in your knowledge and hard work.

Chapter 4

God loves us all and wants us to be with him. Go to him he has the code for he made you. Jesus is the son of God and he came to earth and put on flesh. He died on the cross and was buried. He rose on the third day and ascended to his father where he sits at his right hand interceding for us. God loves his children.

Jesus called me and I am happy I answered his call. Today I pray that I finish the race and get home to my savior. This world tries to make me dizzy with their programs. Sometimes too tire to think straight. When that happens I go on bended knees to my heavenly father who is always near. He guides me and show me the way home.

God loves you and he wants you to spend forever with him. Jehovah gives us free will. The will to choose is powerful. Go to him and he will lead you. How to get save is outline on the last page of this book. Two powerful prays are on the same page under words of encouragement. The choice is yours, choose wisely.

Every day is a chance to repent and go on bended knees and ask God for forgiveness of sins. Once save try not to press the restart button oven and over. Keep going forward to a higher calling. One can ask for forgiveness and turn away from their sins if this is needed however, one has to be careful. Do not get stuck at the starting point because of guilt and shame.

Poem: God Is Love

God is love

He is love and he loves me

God is love

He is love and he cares for me

He meets me at every need I don't want no more

God is love

He is love and he loves me

He meets at every need I don't walk no more

God is love he is love and he carries me

He meets at every need I don't walk no more

God is love

He is love and he loves me

He meets me at every need I don't want no more

God is love

He is love and he loves me

Poem: Set Free

Flying high in the sky

My wings are working

My smile is bright

Keep going

Strong as an eagle

Set free

Poem: His Grace

His grace is sufficient for me

Peace, Love, Faith

Daily grinds

Keep calm

My Secret Heart

God is love

He calls

He waits

He chases after me

He Is a Jealous God

Poem: Rejoice

Thank you Jesus for saving my soul and having compassion on me

Thank you for your loving kindness

Your loving kindness is new every morning

I am grateful

All of my praises belongs to you

Hallelujah….

You are awesome in all your ways

Hallelujah to my savior

Hallelujah to my king

Hallelujah to my healer

Hallelujah to my redeemer

You are awesome in all your ways

JESUS

Priceless

Nothing can compare

You are wonderfully made

Hang in there

I prepare a place for you

Beauty and love lives there

Everything you can imagine that you love and hold dear

Priceless

<u>Courage</u>

You are strong

I know you are trying your best

I will always support you

You are courageous

Be gentle

Love

Peace

Peace I give to you

Breathe

Relax

Don't worry about anything

Don't be anxious

I love you

Peace

Journey

Live your best life

Be wise

Enjoy the fruits of your labor

Don't wish to be here

Lessons to learn along the way

I placed you in time

I am always near

Jesus

My Love

My Love I am here

I hear and see you

Don't give up

Be patient with yourself

Everything takes time

Always with you

I laugh when you laugh

I cry when you cry

I rejoice when you are happy

You are mine

Poem: Remember Me

Dear Lord remember me

Remember me for I am needy

I am depending on you

My heart is heavy

My mind is weary

I feel far away from you

Parachute gives out

Dear Lord remember me

For I am needy

Poem: To Become...

Rising early sleeping late or not all

High fives all around it's graduation and it is fall

Some days running and hide not all make it teary eye

Working with a smile celebrating the good times

Have faith

Stay strong you have to stay strong

Wishing you well even if I don't feel like it

Stay strong tomorrow is a new day

Rising early sleeping late or not at all

To become....

"Your word is a lamp to guide my feet and a light for my path."
Ps.119:105

Words of Encouragement

Life is beautiful and God loves you. He wants you to spend forever with him. Jehovah gives us free will. The will to choose is powerful however; it has to be use in a careful manner. Go to him and he will lead you. The sinners pray and the our father's pray are powerful, both are below:

Father in the name of Jesus I come to you asking you to forgive me for my sins. I have sinned against you and I am sorry. Please forgive me. I humbly pray that you be the lord and savior of my life, Live within me and guide me. In Jesus name, amen.

Our father who art in heaven, hallowed be thy name. Thy kingdom come. Thy will be done in earth, as it is in heaven. Give us this day our daily bread. And forgive us our debts, and we forgive our debtors. And lead us not into temptation, but deliver us from evil: For thine is the kingdom, and the power, and the glory, for ever. Amen. (Matt 6:9-13 KJV)

How to get Save

After praying the above prays and believing them. Give your heart to Jesus. Accept him as your lord and savior. Read your bible. Attend a bible based church. After attending bible study, baptized in full submersion of water in the Father, Son and Holy Ghost.

Congratulations and welcome to the body of Christ

NOTES

NOTES

www.ingramcontent.com/pod-product-compliance
Lightning Source LLC
Chambersburg PA
CBHW070947200526
45161CB00001BA/20